YOUR KNOWLEDGE HAS VALUE

Feasibility Experiments for the Development of Innovative Transdermal Systems

A Combination with Microneedles

Sebastian Kerski

Bibliographic information published by the German National Library:

The German National Library lists this publication in the National Bibliography; detailed bibliographic data are available on the Internet at http://dnb.dnb.de.

ISBN: 9783346286536
This book is also available as an ebook.

© GRIN Publishing GmbH
Nymphenburger Straße 86
80636 München

Print and binding: Books on Demand GmbH, Norderstedt, Germany
Printed on acid-free paper from responsible sources.

The present work has been carefully prepared. Nevertheless, authors and publishers do not incur liability for the correctness of information, notes, links and advice as well as any printing errors.

GRIN web shop: https://www.grin.com/document/946557

Bachelor thesis

Applied Science - Chemistry

Feasibility experiments for the development of innovative transdermal systems in combination with microneedles

author: Sebastian Kerski

ZUYD University of Applied Science, the Netherlands

in cooperation with tesa Labtec Gmbh, Germany

15th January 2015

Bachelor thesis

Applied Science - Chemistry

Feasibility experiments for the development of innovative transdermal systems in combination with microneedles

author: Sebastian Kerski

ZUYD University of Applied Science, the Netherlands in cooperation with tesa Labtec GmbH, Germany

15th January 2015

Acknowledgements

This research was supported by tesa Labtec GmbH in Langenfeld.

Firstly, I would like to express my sincere thanks to my academic supervisor Dr. Evert Vanecht and my business supervisor Dr. Armin Breitenbach for their support of my BA-Thesis.

My appreciation also goes to Dr. Frank Fischer (Beiersdorf AG) and Dr. Nicolai Böhm (tesa SE), who provided me an opportunity to join their team and gave access to the laboratory and research facilities. Without their precious support it would not have been possible to conduct this research.

I thank Sonja Pagel-Wolff, Daniel Mellem and Thomas Lange from Beiersdorf AG, who provided insight and expertise that greatly assisted the research regarding the microscopic technology, Christiane Uhl from Courage+Khazaka electronic GmbH, who provided insight and expertise concerning the TEWL-Technology, and Sebastian Schmidt-Lehr from tesa SE, who provided insight and expertise regarding Micro-CT-technology.

I thank Sandra Lindert, Gabriele Stodt and Marion Tegelkamp for the stimulating discussions.

Last but not least, I would like to thank Eva Kerski for supporting me during this thesis.

Abstract

Transdermal therapeutic systems (TTS) are dosage forms developed to transport an active pharmaceutical ingredient (API) through the skin. This is done by applying a patch to the skin and is therefore a pain-free method. Since the establishment of this method, TTS have been a good alternative to traditional dosage forms such as tablets, injection needles and suppositories.

However, there is a crucial limit to TTS. APIs which are significantly greater than 500Da (500g/mol) cannot pass through the skin barrier. As several publications (cf. Cormier et al., 2004, Roxhed, 2007 and Yu et al., 2015) have shown, there is a possibility to transport APIs which are greater than 500Da through the skin. Combining TTS with microneedles (MN) is a way to produce microchannels through which the API can then pass the skin barrier. The aim of this work was to test the *in-vitro* permeability of APIs across perforated and unperforated skin. Both, traditional adhesive matrices and hydrogel matrices were developed and tested. The result of this work is that it is possible to deliver model substances greater than 500Da through the skin barrier by perforating the epidermis. Customary MN systems were not sufficient to perforate the skin barrier. Hypodermic needles, however, are suitable to perforate the stratum corneum (SC). Consequently, the combination of TTS technology with MN technology is possible and should be further developed. Nevertheless, suitable MN have to be found. Hollow needles which are incorporated into a hydrogel matrix are a very promising option for a future product.

Contents

1. List of Figures

All figures are my own work. They were either plotted using the respective programme or MS

Office 2007 and 2010.

2. List of Tables

3. List of abbreviations

5D-IVT	5D-Intravitaltomograph
API	Active Pharmaceutical Ingredient
cm^2	Square Centimeters
CT	Computed Tomography
CV	Permeationcell Volume
Da	Dalton
DES	Desmopressin Acetate
DEX	Fluorescein Isothiocyanate-Dextran
EMA	European Medicines Agency
h	Hours
HPLC	High Pressure Liquid Chromatograph
hyd	Hydrogel adhesive
IVSP	*In-Vitro* Skin Permeation
μg	Microgram
MCS	Measured Concentration of the Sample

min	Minutes
MN	Microneedles
MYM Stamp	MYM derma stamp pen
OECD	Organisation for Economic Cooperation and Development
PA	Permeation Area
Perm.	Permeation
R^2	Coefficient of determination
SIL	Silicone adhesive
SKYSCAN	SKYSCAN® 1272 High-Resolution X-Ray Microtomograph
SV	Sampling Volume
SC	Stratum Corneum
t	Sampling Point
tL	tesa Labtec GmbH
TTS	Transdermal therapeutic system
TEWL	Transepidermal Water Loss
Vivascope	Vivascope® 1500 confocal laser scanning microscope

4. Introduction

This bachelor thesis deals with the development of a new system for Transfilm®, the TTS technology of tesa Labtec (tL). TTS are drug delivery systems that are applied directly to the skin. The active substance is absorbed by the skin and distributed through the body via the bloodstream. The advantage of Transfilm® is that it allows a safe, reliable, precise and pain-free application with fewer side effects (cf. Gronbach, A., & Kerski, S., 2012). Thus, it becomes easier to treat children, as well as elderly patients and patients requiring complex care. The fact that the transdermal patch can provide a controlled release of medication for up to seven days gives it a major advantage over other types of drug delivery.

To date, there are ca. 20 APIs for which TTS technology has been established (cf. Roxhed, 2007, p. 6). However, many APIs cannot pass the epidermis because of their size. In order to increase the permeability of APIs, there have been many approaches.

8

One of the greatest challenges for tL is the development of new transdermal systems. An innovative possibility is the combination of TTS with MN. The main motivation to use MN is that systemic absorption can be achieved under minimally invasive conditions in cases where regular passive diffusion technologies are not effective enough.

The aim of this paper is to present first experiments for an innovative development of TTS in combination with MN in order to discern the prospects of further projects. The aim is not to define a ready-for-market formula. First, a suitable MN technology needed to be found to make sure that the SC is perforated while the dermis stays intact. Furthermore, TTS formulations were tested to detect how much API is released through the microchannels from a hydrogel matrix or a silicone matrix in order to determine the matrix system that is particularly suited for this technology.

The next chapter describes the theoretical background of the tests. This chapter gives relevant information on the structure of the human skin and TTS and MN technologies. Chapter 6 contains an overview of the used materials and the methods that were employed to test the perforation of the skin and the API release from different matrices. The following chapter (7) comprises statistical analyses of the test results. After that, the results are presented in chapter 8. These are discussed in the final chapter 9.

5. Theoretical background

5.1   The barrier of the skin and transdermal therapeutic systems

This paper primarily deals with Transfilm®, the TTS of tL. TTS are drug delivery systems that are applied directly to the skin. To understand the mechanism of a TTS, it is important to know the structure of the skin. The SC is the barrier of the skin and protects the underlying tissues from heat, microbes and chemicals. Below the epidermis there are basal layers, which serve epidermal wound healing and form a new SC through keratinization (cf. Tortora & Derrickson, 2006, p. 191).

According to recent research the dermis does not function as a barrier for APIs or any other substances. For this reason only the barrier properties of the SC are considered in the context of this thesis. Figure 1 shows a simplified sketch of the skin (cf. Tortora &

9

Derrickson, 2006 p. 191 & 193). The mechanical resistance of the epidermal barrier is mainly due to the corneocytes embedded in the so called cornified envelope. The main biochemical components of the skin barrier are lipids and proteins (cf. Darlenski et al., 2011, p.37).

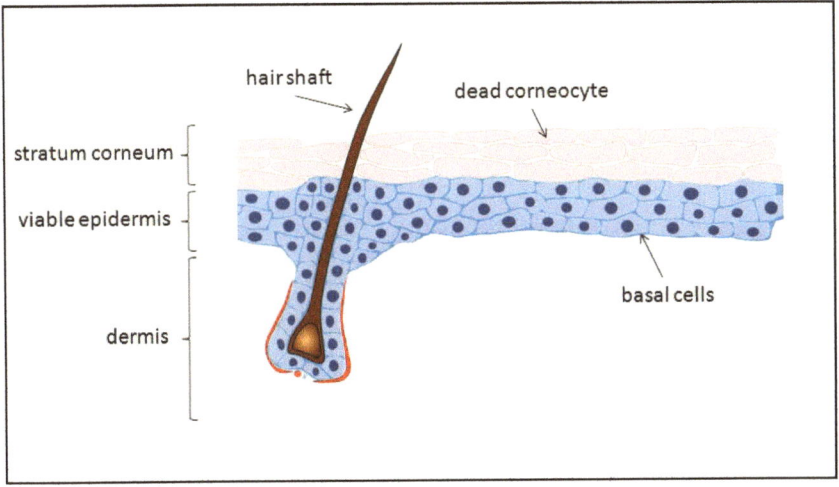

Figure 1: SC with its main constituents, the dead corneocytes.

The API is absorbed by the skin and distributed through the body via the bloodstream. The advantage of the Transfilm® is that it allows a safe, reliable, precise and pain-free application providing fewer side effects (cf. Gronbach, A., & Kerski, S., 2012, p. 1). Thus, it becomes easier to treat children, as well as elderly patients and patients requiring complex care. The fact that the TTS provide a controlled release of medication for up to seven days gives it a major advantage over other drug delivery systems. A well-known TTS is the nicotine patch. Another product which is already on the market is a patch containing fentanyl, an API used in the treatment of chronic cancer pain (Fentanyl Ratiopharm). The fentanyl patch was developed and has always been manufactured by tL (Kerski et al, 2009, p.7). Other cases in which TTS are used are Morbus Parkinson, Morbus Alzheimer and Angina Pectoris (cf. Sharma, R., Puri, D., Bhandari, A., Soni, B., & Singh, M. 2011).

The API should not exceed a certain size (approx. 500Da) so that it can penetrate the skin effectively (cf. Bos, J. D., & Meinardi, M. M., 2000). Furthermore, the active substance should be lipophilic enough to penetrate the skin, but at the same time has to be hydrophilic enough to get into the bloodstream. There is a distinction between three different ways of permeation of APIs through the skin. APIs do not only pass through pores and hair follicles (follicular), but also through microscopic intercellular spaces or through the cells themselves (intracellular).

All three routes are displayed graphically in Figure 2.

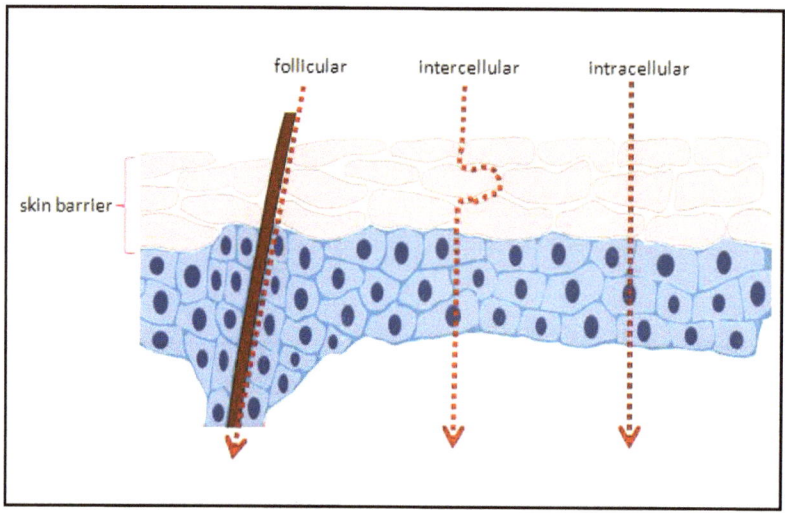

Figure 2: conceptual drawing of different ways of permeation of the APIs through the skin.

On the intercellular route, the API migrates through a lipid matrix which is located between the corneocytes (cf. Chien, Y. W., 1993, p. 132). This is only possible if the API has lipophilic characteristics. To date, there are 20 active substances known to fulfill these requirements (cf. Roxhed, 2007, p. 6). Many APIs cannot pass the epidermis because of their size. Large molecules like peptides or proteins, particularly insulin or vaccines, are therefore not suited to be applied with a patch so far. In order to increase the permeability of drugs many approaches have been described (cf. Barry, 1993, p.119 and Chien, 1993). All of them have tried to enhance the permeability of the SC. For example, chemical enhancers reversibly disrupt the SC structure.

Different APIs require different designs. In general, one can distinguish between two types of transdermal patches (see Figure 3). The matrix system (A) is a semisolid adhesive matrix, a solution or suspension containing an API. The reservoir TTS (B) has a separate API layer. The API layer is a liquid compartment containing a drug solution or suspension separated from the adhesive layer by a membrane. Both types of patches deliver a specific dose of API through the skin and into the bloodstream (cf. Sharma, R., Puri, D., Bhandari, A., Soni, B., & Singh, M., 2011). Figure 3 shows the different types of TTS (cf. Kalvimoorthi, Rajasekaran, Rajan, Balasubramani, & Kumar).

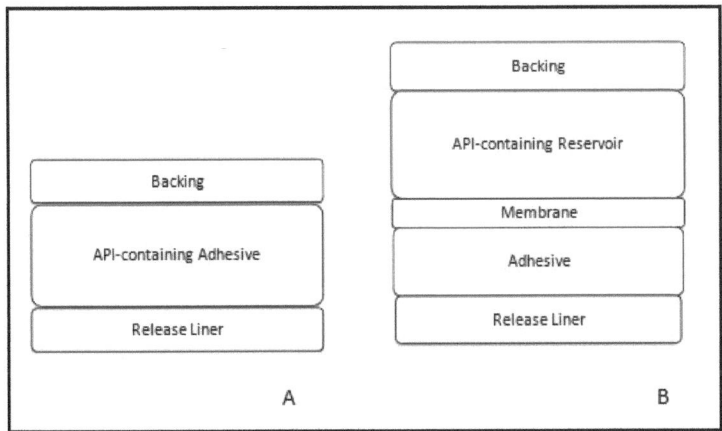

Figure 3: conceptual drawing of a matrix TTS (A) and a reservoir TTS (B).

5.2 Transdermal therapeutic systems in combination with microneedles

One of the greatest challenges for tL is the development of new TTS. One innovative possibility is the combination of TTS and MN. MN are long enough to penetrate the outer layer of the skin but too short to irritate nerves and blood vessels. They can be used to deliver APIs with larger molecules into the skin, an application that is largely pain-free. MN create tiny holes in the outermost layer of the skin (cf. Donnelly et al., 2010, p. 337), which significantly increases the rate at which the active substance is absorbed (cf. Kim et al., 2012, p. 1562). The main motivation to use MN is that systemic absorption can be achieved under minimally invasive conditions in cases where regular passive diffusion technologies are not effective enough. So far, numerous commercial MN systems are available, which are used in medical or cosmetic contexts. The first official product on the market was the so called

Dermaroller® (cf. mi.to.pharm-GmbH, 2015). According to Roxhed (cf. Roxhed, 2007, p. 23) the first publication on solid MN was delivered by Dizon, Han, Russell, & Reed (1993) and the first paper about hollow MN was submitted by Mc Allister et al. (1999). MN can be used to inject active substances with small or large molecules into the skin (cf. Badran, Kuntsche, & Fahr, 2009). The first patent of a TTS combined with MN was published by Gerstel (cf. Gerstel & Place, 1976).

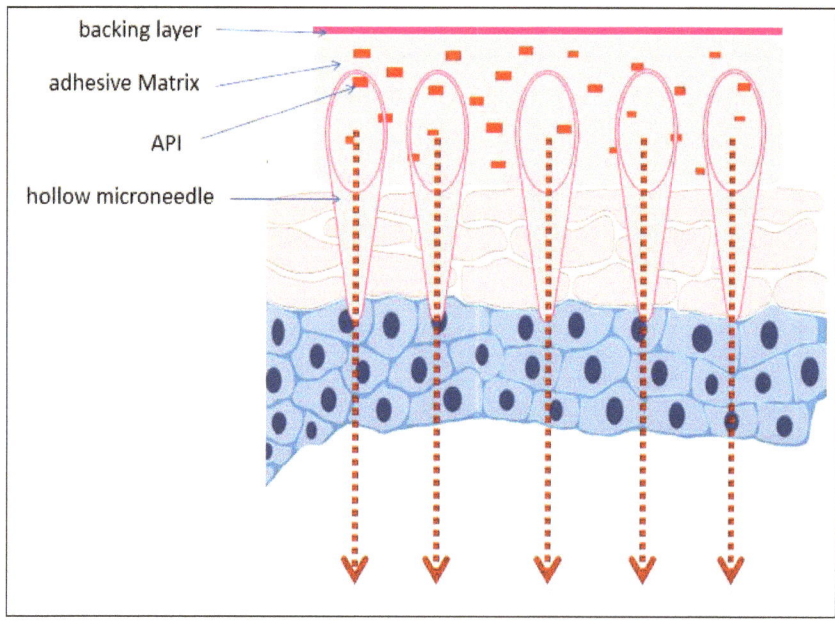

Figure 4: conceptual drawing of a microneedle-based drug delivery system applied to the surface of the skin.

While the application is largely pain-free, it still causes minimal reversible injuries of the epidermis resulting in the exudation of hydrophilic fluids. Traditional matrix systems, however, usually contain lipophilic adhesive matrices based on, for example, acrylate, silicone or styrene. A hydrophilic matrix system might be better suited for a combination of TTS with MN, because in this case the matrix as well as the exuded fluids are hydrophilic. Beiersdorf AG already offers various wound plasters under the name of Hansaplast® and also holds patents on a hydrogel matrix system, which is suitable for transdermal patches (cf.

13

Wöller, 2013).[1]

5.3 *In-vitro* skin permeation tests

In-vivo experiments are the best model to test whether the API can permeate through the skin barrier. As an alternative to this method, this thesis contains *in-vitro* skin permeation tests (IVSP) in an artificial environment. The term *in-vitro* (from Latin; vitrum = glass) means the transfer of experiments outside of the organism, for example into test tubes. The term *ex-vivo* (from Latin; vivus = living) is also used for skin permeation experiments as they are not conducted on living organisms. This term highlights the collection of material from a living organism.

According to the guidelines by the European Medicines Agency (EMA) (2014) and the Organisation for Economic Cooperation and Development (OECD) (2004), IVSP-studies are not expected to parallel *in-vivo* release. IVSP-studies ensure the comparability of an original TTS with development batches that were characterized during the development process and reflect the thermodynamic activity of the API. As part of this thesis, IVSP-tests were performed. The IVSP is a simple and often used test during TTS development. The experiments were performed using a permeation-cell developed at tL (cf. Kerski et al., 2015) which is a further modification of the permeation-cell by Franz (cf. Franz, 1975). The Franz cell consists of a donor compartment in which solutions or patches can be applied to the skin surface. Underneath, there is an acceptor compartment filled with an aqueous buffer. The active substance diffuses from the donor chamber through the skin into the acceptor chamber. Via the sampling port, the sample is taken in regular intervals and replaced by fresh medium. The API concentration in the sample is measured by HPLC.

This makes it possible to trace the permeation of the API through the skin throughout the selected period of time. The temperature stays constant at 32°C (physiological temperature of the skin) during this period (cf. EMA, 2014, p. 25 and OECD, 2004, p. 4). The IVSP is an

---

[1]   Hydrogel matrix systems for the administering of APIs are traditionally used in Asia, particularly in Japan. The Japanese Pharmacopoeia defines the term "Cataplasma" (poultice).

14

important method in dosage form approval. For this reason, the EMA added Appendix 1 to the Guideline on quality of transdermal patches in 2014 (cf. EMA, 2014). The IVSP-tests in this thesis were performed according to the guidelines of EMA and OECD. According to the OECD, the principal diffusion barrier for the API is the non-viable SC (cf. OECD, 2004, p. 1). Damage to the epidermis can be very well observed visually and evaluated. A sampling longer than 24 hours is allowed if the sampling frequency of the receptor fluid allows the absorption profile of the test substance to be presented graphically (cf. OECD, 2004 p. 4).

6. Materials and methods

6.1 Chemicals and reagents

Glycerol 85% Ph. Eur 7.0 and Agar Agar PLV. Ph. Eur 7.0 were purchased from Caesar & Lorenz (Hilden, Germany). Sodium hydroxyde purum 98%, Fluorescein isothiocyanate-dextran average mol wt 4,000, Citric acid monohydrate puriss pa Ph. Eur., Sodiumphosphat puriss pa Ph. Eur, Potassium Phosphate monobasic pa ACS Ph. Eur and Desmopressine European Pharmacopoeia were obtained from Sigma Aldrich (Steinheim, Germany). Carbopol 971 PNF polymer was purchased from Cabrizol Advances (Brussels, Belgium). Purified Water was prepared by filtration of osmosis using the Milli-Q-system from Millipore Advantage A 10 (Darmstadt, Germany). BIO-PSA 7-4201 silicone adhesives were obtained from Dow Corning GmbH (Wiesbaden, Germany). As a liner for the patch preparation Silphan Liner Siliconature SpA (Godega di Sant'Urbano, Italy), HOSTAPHAN® RNT 23 Liner Mitsubishi Polyester Film GmbH (Wiesbaden, Germany) and 3M™ 9733 Scotchpak Backing liner 3M Germany (Neuss, Germany) were used.
Perforation tests were done with Neolus hyperdermic Needle 0.45 x 12mm, Terumo Europe NV (Leuven, Netherlands) and the MYM derma stamp Electric Pen, MYM-Microneedle-Skincare (Fremont, USA). API content in TTS, saturated solution and acceptor solution were analyzed by HPLC Dionex Summit HPLC system consisting of SOR-100 Solvent degasser, P680A HPG-2 High-Pressure Gradient Pump with ASI-100T ™ Automated Sample Injector with 100 ul syringe, TCC Thermostatted Column Compartment, UVD 170U HPLC UV-Vis Detector, Chromeleon® Chromatography Management Software (v. 6.60) (all from Thermo

Fisher Scientific Inc. (Dreieich, Germany)). In addition, Inertsil ODS-2 250 x 4.6 mm from MZ-analysis technology (Mainz, Germany) and the Yarra 3 microns SEC-4000 CC Column 300 * 4.6 mm Phenomenex (Aschaffenburg, Germany) HPLC-columns were used. Additionally, a RID-20A Refractive Index Detector, Shimadzu Europe (Duisburg, Germany) was used.

6.2 Methods

6.2.1 API coating

Manufacturing hydrogel TTS according to the patent (cf. Wöller, 2013), Example IV: A hydrogel was prepared. Initially, there was 40g water. Carbopol and glycerin were dispersed. The sodium hydroxide was dissolved in the remaining water and the agar agar was dispersed in this solution. The two suspensions were then mixed. After a homogeneous gel had formed, the API was added and stirred in the matrix for 3 hours.

Manufacturing silicone TTS: When preparing the silicone TTS, the content of the solid of the adhesive was measured and adjusted adding heptane. After this, the API was added and stirred in the matrix for 3 hours.

Both adhesive matrices were cast on a release liner with a thickness of 300µm using a squeegee, the Quadruple Film Applicator by ERICHSEN GmbH & Co (Hemer, Germany). The TTS were dried for 45min, the hydrogel TTS at 30°C and the silicone TTS at 70°C. After drying, the release liner was laminated. The final system had a total area of 0.82cm² after it was stamped out with a hollow punch.

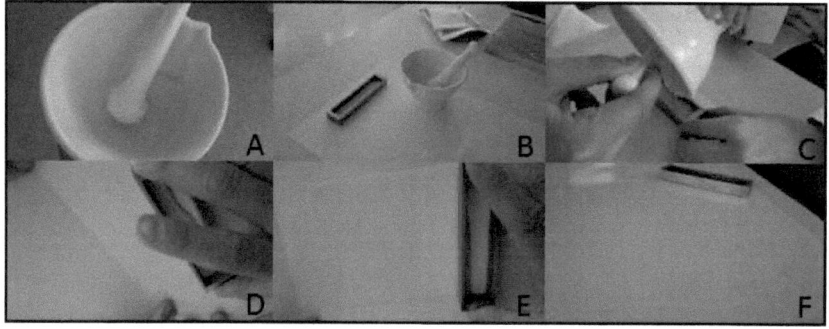

Figure 5: images of the preparation of a TTS, (A) preparing of the hydrogel, (B), (C), (D) & (E) distribution by squeegee, (F) distributed hydrogel matrix.

6.2.2 Skin preparation

Depending on the form of the test, human full thickness skin, heat separated epidermis or split thickness skin was used. The human skin, from female donors, used for the analysis was supplied from plastic surgery. After it had been delivered, the skin was visually checked for scars and stretch marks. Female skin has less hair and follicles than male skin and tL has positive experiences with female skin. The procedure is well established and standardized at tL. The skin used in this test was from the donors' breast or abdomen. Full thickness skin (900 - 1100μm) was prepared using a scalpel, by dissecting the adipose tissue. It was ensured that no contamination of the upper area of the skin had taken place by disserting the adipose tissue (for the method used cf. Kligman & Christophers, 1963). Split thickness skin (200 – 400μm) was cut from the rest of the skin by using a Padgett Dermatome Model S400 from Integra NeuroSciences GmbH (Ratingen, Germany). The dermatome was held at an angle between 25° and 45° and moved forward at a constant velocity. The layer should have a thickness of 200-400μm (cf. OECD, 2004). In order to separate the epidermis from the dermis, the heat separation method was used. For this, the skin sample was put into warm water at the temperature of 60°C for 90 seconds and removed afterwards (cf. Poumay & Coquette, 2007).

6.2.3 Skin perforation

To perforate the skin, derma stamps and hypodermic needles were tested. Each perforation tool was imprinted on the skin with strong pressure. When perforation was done using hypodermic needles, the skin was fixed with an open glass column. The derma stamps were pressed onto the skin without fixing the skin. The microstamp used in the test was the MYM derma stamp pen, which is an advanced version of a microneedling device. The injection depth of the MN can be varied from 0.25mm to 2.0mm (cf. MYM-Microneedle-Skincare, 2015). The injection depth of the hypodermic needles cannot be varied. It only penetrated the epidermis and higher-lying areas of the dermis. The lower layers of the dermis remained imperforated. Figure 6 shows the implementation of the perforation.

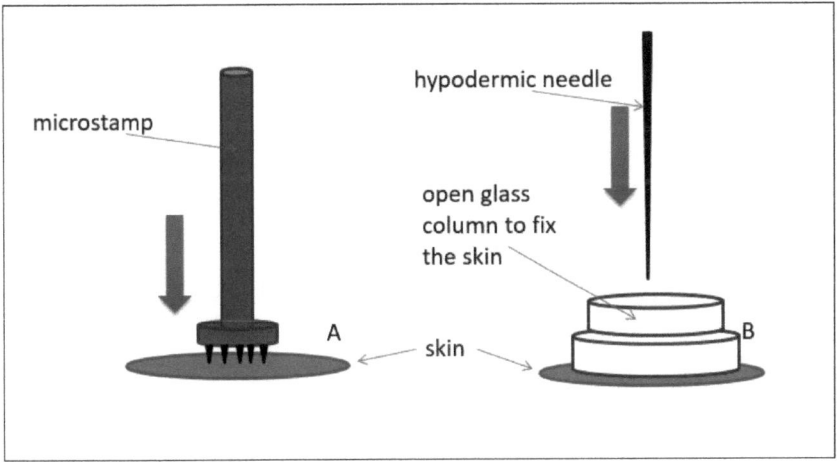

Figure 6: conceptual drawing of the implementation of the perforation.

6.2.4 Verifying the integrity and perforation of the skin

In addition to a visual check at least before and after the perforation, TEER, TEWL, microscopy and micro-computed tomography were performed. The transepidermal water loss (TEWL) was measured using a Tewameter® VT 310. This test was performed at Courage + Khazaka electronic GmbH in Cologne. TEWL-measurement is a highly sensitive method to determine barrier function impairment of the SC. The technology with the open chamber of the Tewameter® is the only one that can perform continuous measurements and record the TEWL value without affecting the microclimate on the skin surface. A large number of studies on this measuring principle are published. For example Kanikkannan et al. (2001) & Netzlaff et al. (2006). Furthermore, in the current EMA Guideline "Guideline on Quality of transdermal patches" in Annex 2 (in-vitro permeation studies) TEWL is evaluated as a suitable method to check the skin integrity (cf. EMA, 2014, p. 25).

The multiphoton tomography was executed using a 5D-IVT. This test was performed at Beiersdorf AG in Hamburg. The basic technology of 5D-IVT is the multiphoton tomography, which shows individual cells and their components in high resolution. The process enables the non-invasive, very detailed presentation of the skin without taking a tissue sample (cf. König et al., 2010). Developed by the company JenLab GmbH, the multiphoton tomograph Dermainspect® is based on the principle of self-fluorescence of living cells and achieves a resolution of less than one micrometer. Selected settings for the measurements can be seen in the following table.

Table 1: Selected settings for the measurements in the 5D-IVT.

Scan Time [s]	Wavelength [nm]	Laser [mW]	Zoom	PMT offset	Gain
7	750	20-40	549	532 PMT	938th

In addition, the laser scanning microscope was used. Laser scanning microscopy was performed using VivaScope® 1500 at the Beiersdorf AG in Hamburg. The confocal laser scanning microscopy shows structures of the skin reaching from the epidermis to the upper reticular dermis non-invasively and in microscopic accuracy. Usually a depth of 200-400µm of the skin can be reached (cf. O'goshi, Suihko, & Serup, 2006).

6.2.5 Penetration assessment

The Penetration depth was measured using the SKYSCAN 1272 at tesa Analytics in Hamburg. Micro-computed tomography is X-ray imaging in 3D. The same method is used in hospital CT scans. Here, it is used on a small scale with a detailed increased resolution. It shows 3D representations of structures on a very fine scale. The internal structure of objects can be determined non-destructively (cf. Bruker, 2015). The skin had to be fixed in place so that it could not slacken in either direction during the measurement. The sample preparation was carried out half an hour before the measurement. The skin was treated with MN and placed on a supporting rigid plastic film (PET 100 microns). The samples were round pieces of skin with a diameter of $0,82cm^2$ to make sure that a resolution of $10-20\mu m$ was reached. The tests were performed on defrosted untreated skin and denatured skin. The defrosted skin was stored for a maximum of 6 months at $-20°C$. The denatured skin was placed into PEG for 1h. Thus, in this procedure, most of the original water content in the skin was replaced by the PEG.

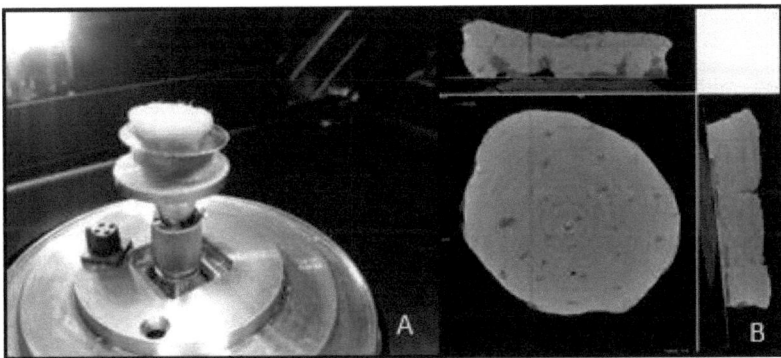

Figure 7: images of *ex-vivo* human skin in the SKYSCAN (A) & after measuring, surface and side view (B).

6.2.6 *In-vitro* skin permeation tests

The *in-vitro* skin permeation is used in this study to examine whether the API passes through the perforated skin. The API incorporated in TTS was applied to the surface of the skin. In pre-tests heat separated epidermis and split thickness skin, and in later tests, exclusively full thickness skin was used. Round pieces ($1.5cm^2$) were stamped out of the skin with a hollow punch. Every tested TTS was evaluated by measuring, six samples per run from two different female donors. For the IVSP a modified permeation cell was used. This cell is protected by the German utility model (cf. Kerski et al., 2015). The static cell is made of glass by Glastechnik Rathsack (Dormagen, Germany). This permeation model is characterized by its permanently filled area inside the acceptor chamber. The acceptor medium permanently stays under the skin. Accordingly, there is no mechanic impact on the skin during the sampling. Additionally, the pressure balance ensures a compensation of the pressure during the sampling. By this modification, very fragile membranes like mucous membranes and heat-separated skin can be tested without a large number of outliers.

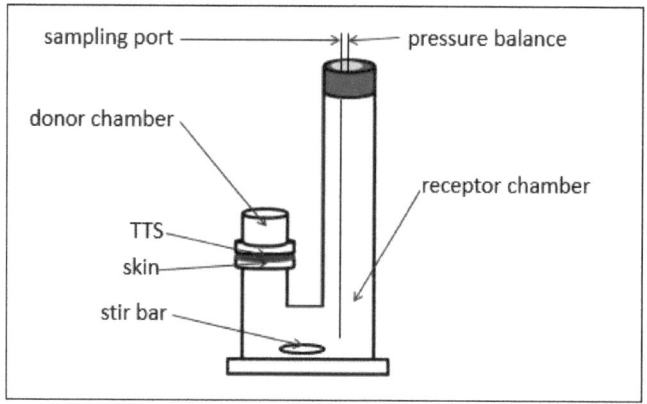

Figure 8: conceptual drawing of the Permeation cell by Kerski, Rathsack and Stodt.

The temperature of the experiment was maintained at $32\pm0.1°C$ using the incubator binder KT115 from Binder GmbH (Tuttlingen, Germany). The temperature control was performed with the ELPRO HAMSTER ET2 data logger from ELPRO Messtechnik GmbH (Stuttgart, Germany). For pressure equalization during sampling, needles for pressure compensation, namely Neolus hyperdermic needles 0.45 x 12 mm by Terumo Europe NV (Leuven, Netherlands) were used.

The medium used for the study was 10mM Citric acid in Milli-Q-Water (acceptor medium). The acceptor medium was chosen to provide sink conditions. To achieve a suitable drug distribution in the acceptor chamber during the permeation, magnetic cylindrical stirring bars, VWR International GmbH (Langenfeld, Germany), and MIX 15 with 15 Multiple Stirrers with integrated Control, 2mag AG (Munich, Germany), were used. The sampling chamber was filled with acceptor medium. The sampling times were 0.5, 1, 3, 6, 9, 12, 18 and

24 hours. At each sampling time, samples of 4.5mL were taken by the Hanson Auto Plus Maximizer System Controller and Hanson AutoPlus MultiFill Fraction Collector purchased from Prosense GmbH (Munich, Germany). Subsamples of the sample solutions were transferred into HPLC vials for analysis. All tests were conducted following the OECD-Guideline and the EMA-Guideline (cf. OECD, 2004 and EMA, 2014). The maximum concentration of the API in the receptor solution achieved during the experiment did not exceed 10% of its maximum solubility in the receptor solution. Visual comparison of chromatograms of the API showed no interference with matrix peaks from the acceptor medium or skin.

The determination of the API in the acceptor medium of the *in-vitro* skin permeation (receptor phase) was done by HPLC with UV-detection according to the established test method. Desmopressin content was measured at 220nm, at a flow rate of 1.0mL/min for isocratic elution on UV spectroscopy. Dextran content was measured at 280nm, at a flow rate of 1.0mL/min for isocratic elution on UV spectroscopy. At the beginning of the analysis, a system suitability test was performed. The calculation of the concentration of the sample was executed by Chromeleon® Chromatography Management Software (v. 6.60) by Thermo Fisher Scientific Inc. (Dreieich, Germany).

The evaluation of the permeation was carried out graphically and by means of the software MS Office Excel 2007. It is important to calculate the volume replacement. This is guaranteed by the following formula. It has to be considered that fresh medium is added while a certain amount of the old medium remains in the cell. The permeation ($\mu g/cm^2/h$) and the steady state flux were calculated by a validated Excel sheet of tL.

Cumulative permeation with results in g/cm^2. Perm = Permeation, MCS = measured concentration of the sample, t = sampling point, PA = permeation area, SV = sample volume and CV = permeationcell volume. The concentration was calculated by the following formula:

$$MCS(t1) \times \frac{VC}{PA} = Perm1$$

$$Perm1 + \frac{MCS(t2) - MCS(t1) \times (CV - SV)}{CV} \times \frac{CV}{PA} = Perm2$$

$$Perm2 + \frac{MCS(t3) - MCS(t2) \times (CV - SV)}{CV} \times \frac{CV}{PA} = Perm3$$

The formulas for Perm4-Perm8 are constructed analogously.

The permeated amount of API was plotted against the experimental period. By dividing the steady state flux J by the API concentration in the patch it is possible to calculate the permeation coefficient P in cm/h, J = API flux [ug/cm²*h] and co = API concentration in the donor [g/cm³].

The permeation coefficient was calculated using the following formula:

$$\frac{J}{co} = P$$

7. Statistical analysis

The 90% confidence interval for the ratio of the TTS was determined and contained within the ratio of 0.8 to 1.25. A Dean-Dixon outlier test was conducted. For this test, the kinetics were compared. A time sequence was chosen in which the deviation of results was most significant. Those results where the deviation was the most extreme were sorted in ascending or descending order. The formula invented by Dean-Dixon was applied (cf. Dean & Dixon, 1951). The result had to be compared with a table by Dean-Dixon. If the result deviated from this table, the results of the tests had to be considered as outliers.

8. Results

8.1 Skin perforation tests

The perforation of the epidermis using micro stamps was impracticable when using the \
MYM Stamp. After puncturing the *in-vivo* skin, an irritation was detected (see Figure 11). These irritations can be detected using Tewameter® (see Figure 4). After four hours, a complete regeneration had taken place.

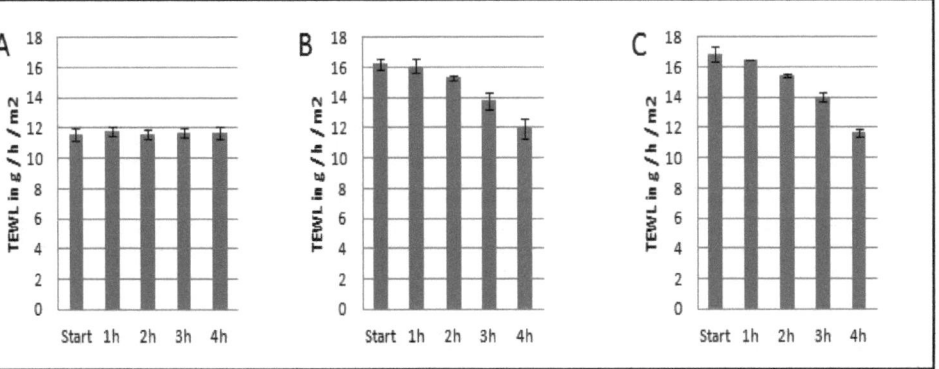

Figure 9: TEWL in g/h/m² *in-vivo* skin, after penetration with "Derma Stamp Electric Pen" Manufacturers MYM. Right arm (B) and left arm (C) vs Reference (A) unperforated skin. Error bars indicate SD (*n* = 2).

Ex-vivo skin was perforated by needles with a penetration depth of 1.5mm and 0.25mm. Here, no difference of TEWL and capacity of skin (by Corneometer®) was detected (see Figures 9 and 10).

24

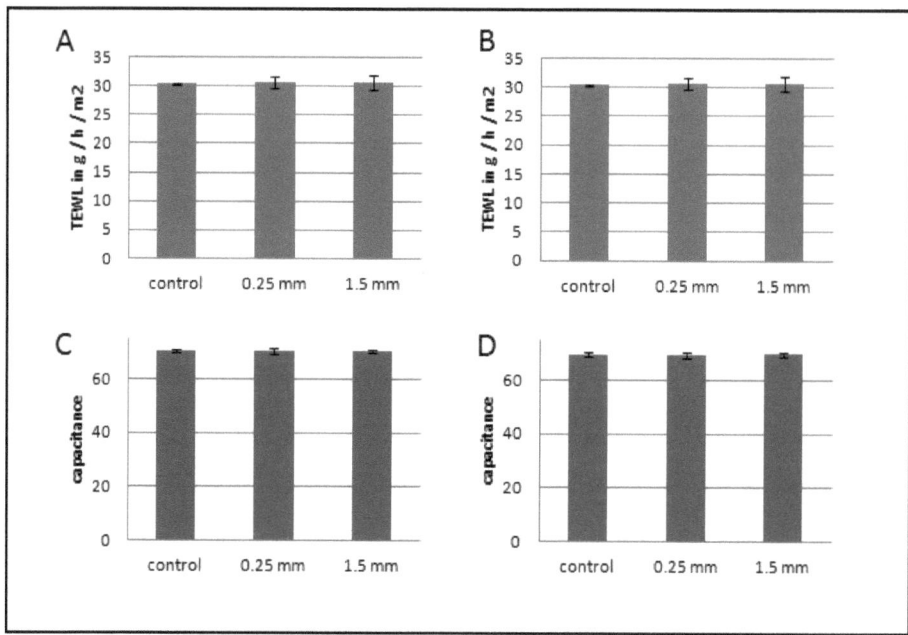

Figure 10: TEWL in g/h/m² and capacitance measurement (Corneometer®) after perforating *ex-vivo* skin when using "Derma Stamp Electric Pen". (A) and (C) full thickness skin, (B) and (D) split thickness skin. Error bars indicate SD (*n* = 6).

The Laser Scanning Microscope (VivaScope®) did not show any damage to the skin. However, an irritation on the surface of the *in-vivo* skin was detected (see Figure 11). The circular marked area designates the treated area. The maximum visual redness could be seen after 3.5h. Using the laser scanning microscope, no effect could be seen over the whole test time of 4h.

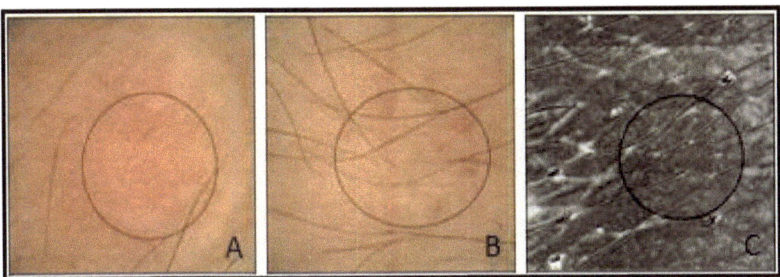

Figure 11: images of (A) freshly treated skin, (B) treated skin after 3.5h, (C) high resolution laser scanning microscope illustration of freshly treated skin.

After attempted perforation of split thickness skin and full thickness skin it was recognized that the MN of the MYM stamp were bent (see figure 12).

Figure 12: images of (A) MN of MYM stamp before perforation, (B) MN of MYM stamp after perforation and (C) hyperdermic needle (no change after peforation).

The measurements using multiphoton microscopy were performed using the system 5D-IVT. Human split skin and full thickness skin were tested. Even with this method no microchannels could be observed. The pictures in Figure 13 (A) - (F) show investigated areas, where no microchannels were detected. These recordings were made over a period of 1min to 25min after injecting MN. The structures found in Figure 13 (G) - (I) were observed, but after

6h no healing process had taken place. Accordingly, these structures were identified as papillae or follicles. The high resolution of the pictures would have made visible any holes made by the MN. The following pictures all show the same skin sample *in-vivo* (human lower arm). (G)-(I) show the same part at different depths.

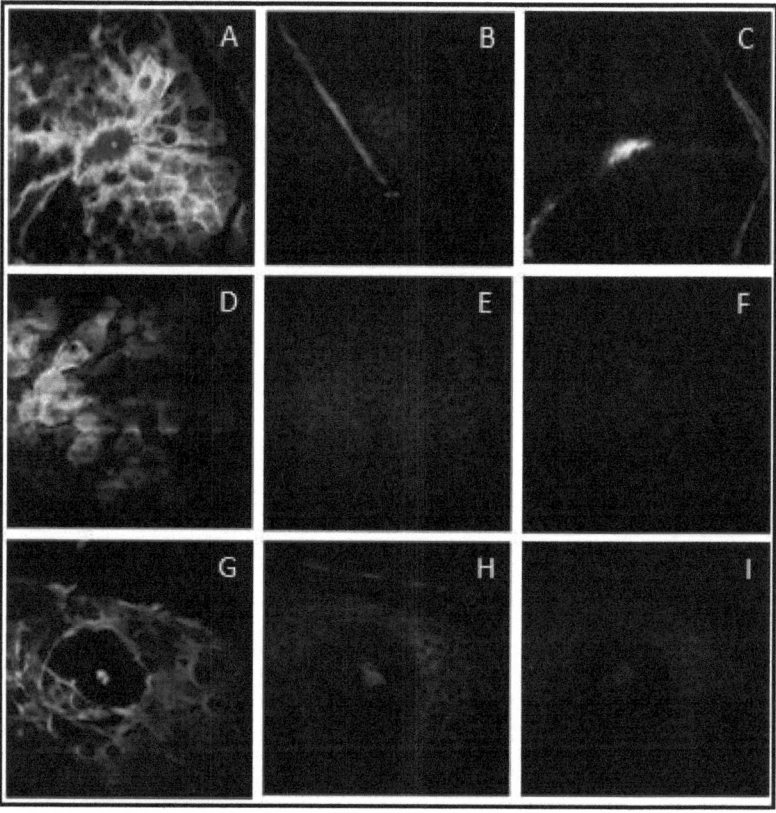

Figure 13: image of *in-vivo* human skin; recording depth (A) 73,9μm, (B) 7,82μm, (C) 16,0μm, (D) 2,0μm, (E) 34μm, (F) 100,0μm, (G) 0,0μm, (H) 28,2μm, (I) 56,0μm.

To confirm the previous results, Micro-CT was used as a further method to analyze the skin. The result was identical. An attempted perforation with MN did not cause any microchannels through the skin.

Pre-tests had shown that the only way to successfully penetrate the skin with MN,

27

using Dermastamps, is to apply these to the epidermis only (heat-separated epidermis). However, the holes need to be visible in unseparated skin. Therefore, other needles had to be used. After fixing of the skin with an open glass column, microchannels were formed with hypodermic needles. Using the SKYSCAN, the microchannels could be made visible in both skins, the defrosted untreated and the denatured full thickness skin (see Figure 14). The untreated skin in Figure 14 dried during measurement and was slightly contracted. The microchannel was 1473μm deep and had a diameter of 154μm at the top. The denatured skin showed the following results: The microchannel in this sample had a depth of 1930μm and a diameter of 562μm at the top.

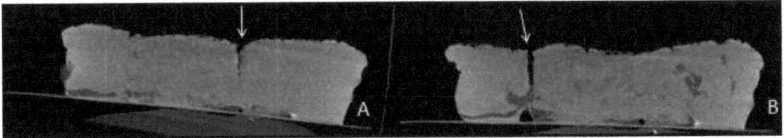

Figure 14: images of *ex-vivo* human skin (A) untreated skin possibly dried and (B) treated skin, before the measurement (water replaced with PEG).

The images demonstrate that only the SC and higher-lying layers of the skin, but not the complete skin was punctured. The perforation in (B) appears to be deeper than in (A). A reason for this observation could be that the untreated skin dried during measurement. This paper will not treat the perforation depth in more detail, because only the perforation of the SC is of importance for the following experiments.

8.2 *In-vitro* skin permeation tests

For the permeation tests, patches with a concentration of 5% API were used. Before the tests started, the concentration was checked several times to demonstrate that the API content of the TTS was stable. Furthermore, a standard was stored in the incubator over the whole duration of the test. After the test of the API the content of the standards was measured, too. The result proved that the solution in the acceptor medium was likewise stable.

Although standard deviations of the experiment are rather low compared to former permeations, it can be concluded that the perforation was executed homogeneously (5 perforations per piece of skin and in the middle of the piece). Flux rates of the hydrogel formulations are higher compared to the silicone formulations. A selective HPLC method for API quantification was established. Furthermore, the saturation solubility of the API at different pH values was determined. The highest solubility was found at pH 5. After ensuring a sufficient stability of the API in this medium, a saturated solution of the API was investigated in a Franz diffusion cell experiment at 32°C for 24h. The permeation kinetics show a lag time of about 3h and a steady-state flux rate after 6h (see Figure 16).

The profiles are rather similar, although slight differences could be detected in steady-state flux rate and SD (see Table 2). However, this variability was expected and is due to the nature of human skin used in the test. The data is summarized in Table 2, Figure 15 and Figure 16. Hydrogel TTS showed a significantly higher performance in the *in-vitro* skin permeation compared to the reference silicone TTS. The reason for this observation could be that the drug can permeate better through the hydrophilic path, coming out of the hydrophilic matrix, while it remains bound by the silicone. The flux rate of the dextran silicone matrix was $0.10\mu g/cm^2/h$. Dextran hydrogel showed a 60 times higher flux rate with $6.16\mu g/cm^2/h$ from 6 to 24h. The correlation coefficient [R^2] for dex silicone TTS was 0.939 from 6 to 24h and for dex hydrogel 0.966. The flux rate of des silicone matrix was $0.20\mu g/cm^2/h$. Desmopressin hydrogel showed a 13 times higher flux rate with $2.6\mu g/cm^2/h$ from 6 to 24h. The correlation coefficient [R^2] for des silicone TTS was 0.982 from 6 to 24h and for des hydrogel 0.981. To ensure that this is an appropriate model, a mass balance was made. This means that the API concentrations in skin and TTS were determined. The masses of API extracted from the TTS and the skin were added to the cumulated amount of API in the acceptor and compared against the API concentration in the TTS. As the API masses are almost identical, this confirms that this model is appropriate. Figure 16 illustrates that the permeation kinetics show a prolonged lag time of about 1h and a steady-state flux rate after 6h. Hydrogel TTS showed a significantly higher performance in the *in-vitro* skin permeation compared to the reference silicone TTS. Error bars indicate SD ($n = 6$).

Table 2: Comparison of the permeation parameters of the four formulations.

	steady state flux [µg/cm²/h] ± SD	permeability coefficient (Kp) [cm/h]	mass API in acceptor after 24 h [µg/cm²] ± SD
Dex hyd TTS	6.16 (0.29)	0,19	124.48 (5.90)
Dex Sil TTS	0.10 (0.01)	0,01	1.78 (0.10)
Des Hyd TTS	2.58 (0.20)	0,41	54.21 (8.86)
Des Sil TTS	0.21 (0.02)	0,01	3.83 (0.26)

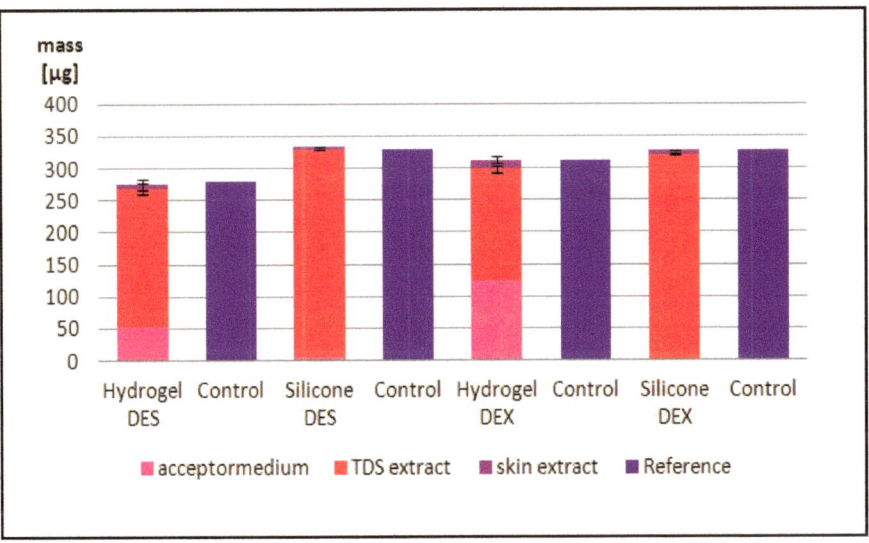

Figure 15: Mass balance of the different formulations of the TTS after IVSP vs. TTS from the assay of the API. The results underline the suitability of the model. Error bars indicate SD ($n = 6$).

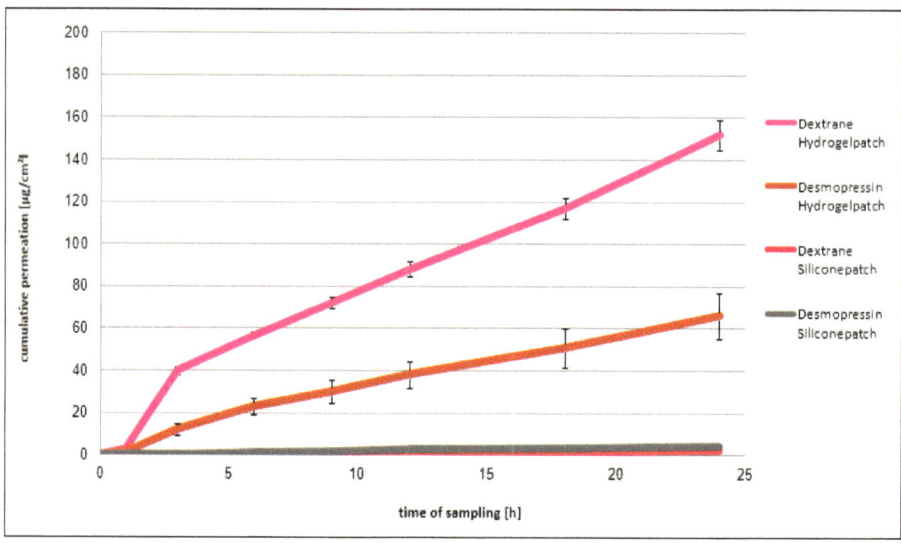

Figure 16: Permeation kinetics. Error bars indicate SD ($n = 6$).

9. Discussion and conclusion

Transdermal delivery of APIs bigger than 500Da across human skin is desirable for drug delivery. There are a variety of publications confirming that the delivery of those APIs on the transdermal way is possible, through the formation of microchannels through perforation by MN. Within this thesis evidence has been produced with regard to the model APIs fluorescein isothiocyanate-dextran and desmopressin-acetate. The desmopressin administration by a TTS in combination with MN has already been shown by Cormier (cf. Cormier et al., 2004). Cormier used needles coated with API, in the same way as Yu et al. (2015). In this case, insulin was used in a so called SmartPatch. For this thesis the technology to produce coated needles was not the focus. Rather, another alternative was meant to be identified. In this work, microchannels formed by perforation were used and TTS types of adhesive matrices were applied. It has been shown that there are significant differences in the API release in the comparative adhesive matrices with the same API loading. This can be attributed to the structure of the adhesive. Both tested APIs showed a higher permeation from the hydrophilic matrix through the micro-channel into the acceptor phase. One possible explanation could be that the polymer chains of silicone bind the large molecules of the API much stronger than the

31

hydrogel. A second factor could be that the dissolution from the hydrogel is easier, because the API permeates from a hydrophilic matrix into another hydrophilic medium.[2] A possible influence of the adhesive properties of the TTS was not taken into account because the skin with the TTS was fixed between a stainless steel grid and the flange. There were several methods used for imaging the skin. For this feasibility study, the measurement showed that the SKYSCAN is the most appropriate method. Measuring the *ex-vivo* skin with the SKYSCAN, the size and depth of the microchannels could be determined. The experiments with the Tewameter, Corneometer, 5D-IVT and Vivascope were able to show the selectivity and limits of the methods. For follow-up experiments measurements with 5D-IVT (*ex-vivo* and *in-vivo*) and SKYSCAN (*in-vitro*) are recommended.

A TTS in combination with MN is an option for a future dosage form for large API molecules. Developing a matrix-TTS containing microneedles is the next step to establish this dosage form. The most practicable form seems, at this moment, to be a matrix system containing hollow MN. Furthermore, the adhesive matrix should be hydrogel-based as the tests have shown that this matrix causes a higher API release. In this case the API-containing matrix would flow into the hollow needles during the coating process. Thus, there would be a direct hydrophilic path into the dermis right after the application of the TTS. With this, there is no delay of the API entering the human organism, because the MN would not really be hollow but already be filled with the matrix substance. Further tests are recommended.

The next challenge for a possible prototype is the application of needles. As my tests have shown, conventional MN cannot perforate the SC, but a certain amount of pressure is required. Several publications suggest the usage of applicators such as Cormier et al. (2004). The most desirable product would be a microneedle patch with an integrated applicator. So far, most of the applicators are not integrated. However, there are patents such as one held by LTS Lohmann Therapie-Systeme AG (Andernach, Germany) (cf. Mohr, 2015) which contain an advancing element. Oval chips can be an option, which are slightly curved and can be

---

[2]   This would also be the case under physiological circumstances as the MN cause exudation of hydrophilic fluids on *in-vivo* skin.

pushed inward and then remain rigid. By pressing the patch to the skin after application, the needles would be pushed into the SC with a defined pressure and thus form a microchannel.

10. Sources

Badran, M., Kuntsche, J., & Fahr, A. (2009). Skin penetration enhancement by a microneedle device (Dermaroller®) in vitro: dependency on needle size and applied formulation. *european journal of pharmaceutical sciences, 36*(4), 511-523.

Barry, B. W. (1993). Skin penetration enhancers-the chemical approach. PAPERBACK APV, 31, 119-119.

Bodde, H., AALTEN, E., & Junginger, H. (1989). Hydrogel Patches for Transdermal Drug Delivery; In-vivo Water Exchange and Skin Compatibility. *Journal of pharmacy and pharmacology, 41*(3), 152-155.

Bos, J. D., & Meinardi, M. M. (2000). The 500 Da rule for the skin penetration of chemical compounds and drugs. *Experimental dermatology, 9*(3), 165-169.

Bruker. (2015). High-resolution micro-CT. Retrieved 29.09.2015, 2015, from https://www.bruker.com/de/products/x-ray-diffraction-and-elemental-analysis/x-ray-micro-ct/SKYSCAN-1272/overview.html

Chien, Y. W. (1993). Systemic delivery of peptide-based pharmaceuticals by transdermal periodic iontotherapeutic system. *PAPERBACK APV, 31*, 129-129.

Cormier, M., Johnson, B., Ameri, M., Nyam, K., Libiran, L., Zhang, D. D., & Daddona, P. (2004). Transdermal delivery of desmopressin using a coated microneedle array patch system. *Journal of controlled release, 97*(3), 503-511.

Darlenski, R., Kazandjieva, J., & Tsankov, N. (2011). Skin barrier function: morphological basis and regulatory mechanisms. *J. Clin. Med, 4*, 36-45.

Dean, R. B., & Dixon, W. J. (1951). Simplified statistics for small numbers of observations. *Analytical Chemistry, 23*(4), 636-638.

Dizon, R., Han, H., Russell, A. G., & Reed, M. L. (1993). An ion milling pattern transfer technique for fabrication of three-dimensional micromechanical structures. *Microelectromechanical Systems, Journal of, 2*(4), 151-159.

Donnelly, R. F., Garland, M. J., Morrow, D. I., Migalska, K., Singh, T. R. R., Majithiya, R., & Woolfson, A. D. (2010). Optical coherence tomography is a valuable tool in the study of the effects of microneedle geometry on skin penetration characteristics and in-skin dissolution. Journal of Controlled Release, 147(3), 333-341.

EMA. (2014). *Guideline on quality of transdermal patches.*

Franz, T. J. (1975). Percutaneous absorption. On the relevance of in vitro data. *Journal of Investigative Dermatology, 64*(3), 190-195.

Gerstel, M. S., & Place, V. A. (1976). Drug delivery device: Google Patents.

Gronbach, A., & Kerski, S. (2012). *Einführung von Projektcontrolling bei der Labtec GmbH*: GRIN Verlag.

Kalvimoorthi, V., Rajasekaran, M., Rajan, V. S., Balasubramani, K., & Kumar, P. S. Transdermal Drug Delivery System: An Overview.

Kanikkannan, N., Burton, S., Patel, R., Jackson, T., Shaik, M. S., & Singh, M. (2001). Percutaneous permeation and skin irritation of JP-8+ 100 jet fuel in a porcine model. *Toxicology letters, 119*(2), 133-142.

Kim, Y. C., Park, J. H., & Prausnitz, M. R. (2012). Microneedles for drug and vaccine delivery. Advanced drug delivery reviews, 64(14), 1547-1568.

Kerski, S., Becker, J., & Gronbach, A. (2009). *Entwicklung eines Analyseverfahrens zur Bestimmung von pharmazeutischen Kreuzkontaminationen an Laborglas und Herstellungsgeräten:* GRIN Verlag.

Kerski, S., Rathsack W., & Stodt G. (2015). Germany Patent No. 20 2015 004 165.5. DPA.

KLIGMAN, A. M., & CHRISTOPHERS, E. (1963). Preparation of isolated sheets of human stratum corneum. *Archives of Dermatology, 88*(6), 702-705.

König, K., Weinigel, M., Breunig, H. G., Gregory, A., Fischer, P., Kellner-Höfer, M., . . . Stracke, F. (2010). *5D-intravital tomography as a novel tool for non-invasive in-vivo analysis of human skin.* Paper presented at the BiOS.

McAllister, D., Cros, F., Davis, S., Matta, L., Prausnitz, M., & Allen, M. (1999). *Three-dimensional hollow microneedle and microtube arrays.* Paper presented at the Transducers.

mi.to.pharm-GmbH. (2015). Der orginal Dermaroller zum Medical Micro Needling. Retrieved 30.08.2015, 2015, from http://www.original-dermaroller.de/de/

Mohr, P. (2015). Transdermales therapeutisches System mit Druckerzeugungsvorrichtung: Google Patents.

MYM-Microneedle-Skincare. (2015). What Exactly is a derma stamp? Retrieved 29.08.2015, from http://mym-microneedle-skincare.com/what-is-derma-stamp-pen/

Netzlaff, F., Kostka, K.-H., Lehr, C.-M., & Schaefer, U. F. (2006). TEWL measurements as a routine method for evaluating the integrity of epidermis sheets in static Franz type diffusion cells in vitro. Limitations shown by transport data testing. *European journal of pharmaceutics and biopharmaceutics, 63*(1), 44-50.

O'goshi, K. i., Suihko, C., & Serup, J. (2006). In vivo imaging of intradermal tattoos by confocal scanning laser microscopy. *Skin research and technology, 12*(2), 94-98.

OECD, T. G. (2004). 428: Skin absorption: in vitro Method. *OECD Guidelines for the Testing of Chemicals, Section, 4.*

Poumay, Y., & Coquette, A. (2007). Modelling the human epidermis in vitro: tools for basic and applied research. *Archives of dermatological research, 298*(8), 361-369.

Roxhed, N. (2007). A fully integrated microneedle-based transdermal drug delivery system.

Sharma, R., Puri, D., Bhandari, A., Soni, B., & Singh, M. (2011). TRANSDERMAL DRUG DELIVERY SYSTEM: AN OVERVIEW. *Inventi Rapid: NDDS.*

Tortora, G. J., & Derrickson, B. H. (2006). *Anatomie und Physiologie:* Wiley-VCH.

WÖLLER, K.-H. U., Katharina. (2013). Deutschland Patent No. DE102010038312A1. EP Register: B. AG.

Yu, J., Zhang, Y., Ye, Y., DiSanto, R., Sun, W., Ranson, (2015). Microneedle-array patches loaded with hypoxia-sensitive vesicles provide fast glucose-responsive insulin delivery. *Proceedings of the National Academy of Sciences, 112*(27), 8260-8265.

YOUR KNOWLEDGE HAS VALUE

- We will publish your bachelor's and
  master's thesis, essays and papers

- Your own eBook and book -
  sold worldwide in all relevant shops

- Earn money with each sale

Upload your text at www.GRIN.com
and publish for free